FLORA OF TROPICAL EAST AFRICA

LILIACEAE

JOHN GRIMSHAW*

Erect perennial geophytes; bulbs formed of 1 to many scales attached to reduced stem, tunicate or not. Leaves basal or cauline, sometimes petiolate, 1- many, alternate or verticillate, linear to ovate-lanceolate, bases sometimes sheathing, veins parallel. Inflorescence a terminal raceme, sometimes umbel-like, or a single terminal flower, bracts subtending flowers absent or present and leaf-like. Flowers hermaphrodite, actinomorphic or occasionally weakly zygomorphic. Perianth segments 6 in 2 whorls, free to base, usually brightly coloured, often with contrasting basal nectaries. Stamens in 2 series of 3; anthers dehiscing by longitudinal slits. Ovary superior, trilocular, with numerous ovules per locule; style solitary, erect, short to long; stigma capitate to 3-lobed. Fruit a loculicidal capsule with numerous seeds. Seeds flattened, discoid to ellipsoidal.

Twelve genera and 350–400 species, all in the Northern Hemisphere, especially south-western and Himalayan Asia to China.

As recently circumscribed (e.g. Dahlgren, Clifford & Yeo, Families of the Monocots: 233 (1985)), Liliaceae is a homogenous family of mostly spring-flowering temperate bulbous plants.

LILIUM

L., Sp. Pl. 1: 302 (1753) & Gen Pl. 5th ed.: 143 (1754); Elwes, Monogr. Genus Lilium 1–83 (1877–1880); Woodcock & Stearn, Lilies of the World: 1–431 (1950); Synge, Lilies: 1–276 (1980); McRae, Lilies: 1–392 (1998)

Perennial herbs with an annual shoot arising from a many-scaled bulb; bulbs made of overlapping, fleshy scales, with strong perennial roots; stem erect, unbranched, leafy, sometimes producing roots at the base. Leaves numerous, verticillate or scattered, usually decreasing in size up the stem, linear or lanceolate, occasionally broader. Inflorescence terminal, a raceme, or with solitary flowers. Flowers pedicellate, subtended by bracts, nodding to erect, funnel-shaped, cup-shaped, bowl-shaped or bell-shaped, often with the perianth segments somewhat or greatly reflexed. Perianth segments 6, petaloid and showy, free, caducous, each with a basal nectary-furrow, sometimes papillose on inner surface, occasionally hairy externally. Anthers large, epipeltate or pseudobasifixed, versatile; filaments in two whorls, inserted at base of perianth segments, filiform to slightly flattened, bearing anthers at the mouth of the flower or greatly exserted from it. Ovary superior, 3-lobed. Style long, stigma 3-lobed (sometimes obscurely so). Fruit a loculicidal capsule, bearing seeds in 2 rows within each locule. Seeds brown, flat but not winged.

About 100 species, mostly in north temperate region, with some species occurring at higher altitudes in tropical and sub-tropical Asia. One species naturalized in East Africa.

*Sycamore Cottage, Colesbourne, nr Cheltenham, Gloucestershire GL53 9NP

FIG. 1. *LILIUM FORMOSANUM* — **1**, habit, × ⅔ (about 15 cm of intermediate stalk not shown); **2**, flower, outer perianth segment and stamen, × ⅔; **3**, flower, inner perianth segment and stamen, × ⅔; 4, style and ovary, × ⅔. 1 from *Coveny* 6107; 2, 3, 4 from *Coveny* 6060. Drawn by Juliet Williamson.

Lilium formosanum *Wallace* in The Garden 40: 442 (1891); Stapf in Bot. Mag. 154: t. 9205 (1930); Price in Gard. Chron. 89: 70 (1931); Grove & Cotton in Suppl. Elwes' Monogr. Genus Lilium 13, t. 5 (1935); Woodcock & Stearn in Lilies of the World: 219 (1950); Synge, Lilies: 80 (1980); McRae, Lilies: 134 (1998). Type: not located.

Herb to 200 cm high, evergreen if not too dry, producing a succession of flowering stems, but in East Africa usually producing a solitary stem during the rainy season. Bulb subglobose, 2–4 cm in diameter, white, tinged purple, stoloniferous. Stems usually solitary per bulb, rooting at the base above the bulb, purple-brown, especially below, scaberulous. Leaves numerous, most dense at base of stem, sparser and shorter above, linear to narrowly oblong-lanceolate, 7.5–20 × (0.2–)0.4–0.5(–1) cm, 3–7 veined, glabrous, lustrous dark green; upper leaves lanceolate, 2–5 cm long, acute. Flowers usually 1(–2) in Africa, potentially 3–10(–30), borne horizontally, white, narrowly funnel-shaped, the outer whorl of perianth segments suffused brownish-pink externally, fragrant; pedicels erect or ascending, 6–15 cm, with leafy bracts and bracteoles. Perianth segments 12.5–20 cm long, reflexed at apices, outer whorl oblong-oblanceolate 2.5–3 cm wide, narrower at base, inner whorl spathulate, with long claw, limb obovate-lanceolate to 5 cm wide; nectary-furrow narrow, green, papillose-pubescent. Filaments equalling perianth tube, 7–12.5 cm long, flattened, obscurely papillose below; anthers 9–21 mm long, more or less included in perianth, pollen brown or yellow. Ovary lustrous green, cylindric, 5 cm long; style angular, 6–9 cm, thickened and curved upwards below the large, 3-lobed stigma, exceeding anthers. Capsule cylindrical, weakly 6-angled, erect, 7–9 × ± 2 cm, apex depressed, base stipitate. Seed obovate, thin with a thickened margin, 5 mm long. Fig. 1.

KENYA. North Nyeri District: Naro Moru, Dec. 1981, *Grey-Wilson* s.n. (cultivated Suffolk from seed collected in Kenya)!

TANZANIA. Lushoto District: Jaegertal Hotel Garden, 18 Oct. 1984, *Kisena* 195!; Iringa District, Kigogo near old foresters's house, 9 Mar. 1987, *Lovett, Keeley & Niblett* 1679!

DISTR. **K** 4; **T** 2, 3, 5, 7; indigenous and naturally endemic to Taiwan, now naturalized in eastern and southern Africa, Australia (where considered a noxious weed)

HAB. Usually semi-disturbed habitats amongst shrubs and grasses, especially by roadsides; ?500–2000 m

SYN. *L. longiflorum* Thunb. var. *formosanum* Baker in Gard. Chron. (n.s) 14: 524 (1880)
L. philippinense Baker var. *formosanum* Grove in Gard. Chron. 70: 63 (1921)

NOTE. Other species of *Lilium* have been, or may be, cultivated in gardens or for cut flowers in East Africa. Among these are the 'white lilies' *Lilium longiflorum* Thunb. (Easter lily) differing from *L. formosanum* in its shorter stature, broader leaves and pure white flowers with scarcely reflexed perianth segments, and *L. candidum* L. (Madonna lily) with broad basal leaves, reduced cauline leaves, and widely spreading pure white perianth segments.

INDEX TO LILIACEAE

No new names validated in this part

PLANTS PEOPLE
POSSIBILITIES

First published in 2005 by
Royal Botanic Gardens, Kew
Richmond, Surrey, TW9 3AB, UK
www.kew.org

ISBN 1 84246 118 4

Design by Media Resources, typesetting and page layout by Margaret Newman,
Information Services Department,
Royal Botanic Gardens, Kew.

Printed by Cromwell Press Ltd.

For information or to purchase all Kew titles please visit
www.kewbooks.com or email publishing@kew.org

ANGIOSPERMAE

Bignoniaceae
Bischofiaceae — in Euphorbiaceae
Bixaceae (£1.50)
*Bombacaceae (£3.90)
*Boraginaceae (£14.80)
Brassicaceae — see Cruciferae
Brexiaceae (£1.50)
Buddlejaceae — as Loganiaceae
*Burmanniaceae (£3.00)
*Burseraceae (£13.30)
Butomaceae (£1.50)
Buxaceae (£1.50)

Cabombaceae (£1.50)
Cactaceae (£1.50)
Caesalpiniaceae — in Leguminosae
Callitrichaceae (£4.00)
Campanulaceae (£4.50)
Canellaceae (£1.50)
Cannabaceae (£1.50)
Cannaceae — with Musaceae
Capparaceae (£7.50)
Caprifoliaceae (£1.50)
Caricaceae (£1.50)
Caryophyllaceae (£3.00)
*Casuarinaceae (£2.00)
Cecropiaceae — with Moraceae
*Celastraceae (£13.00)
*Ceratophyllaceae (£1.50)
Chenopodiaceae (£3.00)
Chrysobalanaceae — as Rosaceae
Clusiaceae — see Guttiferae
Cochlospermaceae (£1.50)
Colchicaceae (£14.50)
Combretaceae (£8.90)
Commelinaceae
Compositae
  *Part 1 (£32.00)
  *Part 2 (£35.00)
  Part 3 (£52.00)
Connaraceae (£3.00)
Convolvulaceae (£13.00)
Cornaceae (£1.50)
Costaceae — as Zingiberaceae
*Crassulaceae (£11.00)
*Cruciferae (£11.00)
Cucurbitaceae (£13.00)
Cyanastraceae — in Tecophilaeaceae
*Cyclocheilaceae (£1.75)
Cymodoceaceae (£4.00)
Cyperaceae
Cyphiaceae — as Lobeliaceae

*Dichapetalaceae (£3.70)
Dilleniaceae (£1.50)
Dioscoreaceae (£3.00)
Dipsacaceae (£3.00)
*Dipterocarpaceae (£3.90)
Dracaenaceae
Droseraceae (£1.50)

*Ebenaceae (£11.00)
Elatinaceae (£1.50)
Ericaceae
*Eriocaulaceae (£6.50)
*Eriospermaceae (£2.00)
*Erythroxylaceae (£3.00)
Escalloniaceae (£1.50)
Euphorbiaceae
  *Part 1 (£31.50)
  *Part 2 (£22.00)

Fabaceae — see Leguminosae
Flacourtiaceae (£5.90)
Flagellariaceae (£1.50)
Fumariaceae (£1.50)

Gentianaceae (£15.00)
Geraniaceae (£3.00)
Gesneriaceae
Gisekiaceae — as Aizoaceae
Goodeniaceae (£1.50)
Gramineae (£74.00)
  Part 1 (£13.00)
  Part 2 (£38.00)
  *Part 3 (£54.50)
Gunneraceae — as Haloragaceae
Guttiferae (£4.50)

Haloragaceae (£3.00)
Hamamelidaceae (£1.50)
*Hernandiaceae (£3.00)
Hippocrateaceae — in Celastraceae
Hugoniaceae — in Linaceae
*Hyacinthaceae (£6.00)
Hydnoraceae (£4.00)
*Hydrocharitaceae (£3.25)
*Hydrophyllaceae (£1.85)
*Hydrostachyaceae (£3.00)
Hymenocardiaceae — with Euphorbiaceae
Hypericaceae (£3.00) — see also Guttiferae
Hypoxidaceae

Icacinaceae (£3.00)
Illecebraceae — as Caryophyllaceae
*Iridaceae (£15.00)
Irvingiaceae — as Ixonanthaceae
*Ixonanthaceae (£3.00)

Juncaceae (£3.00)
Juncaginaceae (£1.50)

Labiatae
Lamiaceae — see Labiatae
*Lauraceae (£4.00)
Lecythidaceae (£1.50)
Leeaceae — with Vitaceae
Leguminosae (£74.00)
  Part 1, Mimosoideae (£13.00)
  Part 2, Caesalpinioideae (£18.50)
  Part 3 }
  Part 4 } Papilionoideae (£59.00)
Lemnaceae (£3.00)
Lentibulariaceae (£3.00)
Liliaceae (*s.s.*) (£12.90)
Limnocharitaceae — as Butomaceae
Linaceae (£3.00)
*Lobeliaceae (£8.30)
Loganiaceae (£4.50)
*Loranthaceae (£12.75)
*Lythraceae (£11.20)

Malpighiaceae (£3.00)
Malvaceae
Marantaceae (£3.00)
Melastomataceae (£9.00)
*Meliaceae (£11.00)
Melianthaceae (£1.50)
Menispermaceae (£3.00)
*Menyanthaceae (£2.00)
Mimosaceae — in Leguminosae
Molluginaceae — as Aizoaceae
Monimiaceae (£1.50)
Montiniaceae (£1.50)
*Moraceae (£14.25)
*Moringaceae (£2.75)
Muntingiaceae — with Tiliaceae
Musaceae (£3.90)
*Myricaceae (£3.00)
*Myristicaceae (£2.70)
*Myrothamnaceae (£1.80)
*Myrsinaceae (£4.30)
*Myrtaceae (£17.00)

ANGIOSPERMAE

*Najadaceae (£3.90)
Nectaropetalaceae — in Erythroxylaceae
*Nesogenaceae (£1.50)
*Nyctaginaceae (£4.00)
*Nymphaeaceae (£3.90)

Ochnaceae (£17.70)
Octoknemaceae — in Olacaceae
Olacaceae (£3.00)
Oleaceae (£3.00)
Oliniaceae (£1.50)
Onagraceae (£3.00)
Opiliaceae (£1.50)
Orchidaceae
  Part 1, Orchideae (£18.50)
  *Part 2, Neottieae, Epidendreae (£22.00)
  *Part 3, Epidendreae, Vandeae (£24.00)
Orobanchaceae (£1.50)
Oxalidaceae (£3.00)

*Palmae (£8.50)
Pandaceae — with Euphorbiaceae
*Pandanaceae (£3.90)
Papaveraceae (£1.50)
Papilionaceae — in Leguminosae
Passifloraceae (£6.00)
Pedaliaceae (£3.00)
Periplocaceae — see Apocynaceae
Phytolaccaceae (£1.50)
*Piperaceae (£5.00)
Pittosporaceae (£3.00)
Plantaginaceae (£1.50)
Plumbaginaceae (£3.00)
Poaceae — see Gramineae
Podostemaceae (£9.80)
Polygalaceae
Polygonaceae (£4.50)
Pontederiaceae (£1.50)
Portulacaceae (£10.00)
Potamogetonaceae
Primulaceae (£3.00)
*Proteaceae (£3.25)
*Ptaeroxylaceae (£2.00)

*Rafflesiaceae (£2.00)
Ranunculaceae (£3.00)
Resedaceae (£1.50)
Restionaceae (£12.90)
Rhamnaceae (£4.50)
Rhizophoraceae (£3.00)
Rosaceae (£6.00)
Rubiaceae
  Part 1 (£22.00)
  *Part 2 (£31.50)
  *Part 3 (£23.70)
*Ruppiaceae (£1.95)
*Rutaceae (£7.75)
*Salicaceae (£2.05)
Salvadoraceae (£3.00)
Santalaceae (£15.00)
*Sapindaceae (£15.00)
Sapotaceae (£7.50)
Scrophulariaceae
Scytopetalaceae (£1.50)
Selaginaceae — in Scrophulariaceae
*Simaroubaceae (£3.40)
*Smilacaceae (£1.85)
Solanaceae
Sonneratiaceae (£1.50)
Sphenocleaceae (£1.50)
Sterculiaceae
Strychnaceae — in Loganiaceae
*Surianaceae (£2.00)

Taccaceae (£1.50)
Tamaricaceae (£1.50)
Tecophilaeaceae (£1.50)
Ternstroemiaceae — in Theaceae
Tetragoniaceae — in Aizoaceae
Theaceae (£1.50)
Thismiaceae — in Burmanniaceae
Thymelaeaceae (£4.50)
*Tiliaceae (£20.50)
Trapaceae (£1.50)
Tribulaceae — in Zygophyllaceae
*Triuridaceae (£1.85)
Turneraceae (£3.00)
Typhaceae (£1.50)

Uapacaceae — in Euphorbiaceae
*Ulmaceae (£3.00)
*Umbelliferae (£14.80)
*Urticaceae (£11.00)

Vacciniaceae — in Ericaceae
Vahliaceae (£1.50)
Valerianaceae (£3.00)
Velloziaceae (£3.00)
*Verbenaceae (£17.80)
*Violaceae (£6.30)
*Viscaceae (£6.20)
*Vitaceae (£17.80)

*Xyridaceae (£6.20)

*Zannichelliaceae (£1.40)
*Zingiberaceae (£5.75)
*Zosteraceae (£1.70)
*Zygophyllaceae (£3.15)

Parts of this Flora, unless otherwise indicated, are obtainable from:
Royal Botanic Gardens, Kew, Richmond, Surrey TW9 3AB, England. www.kew.org or www.kewbooks.com

*** only available through CRC Press at:**
UK and Rest of World (except North and South America): CRS Press/ITPS,
Cheriton House, North Way, Andover, Hants SP10 5BE.
e: uk.tandf@thomsonpublishingservices. co.uk

North and South America:
CRC Press,
2000NW Corporate Blvd, Boco Raton, FL 33431-9868, USA.
e: orders@crcpress.com

ISBN 1-84246-118-4
9 781842 461181

ISBN 1 84246 118 4